Createspace publishing
4900 laCross Road
North Charleston SC, 29406, USA

Published by Createspace: 011/18/2015

ISBN-13: 978-1519370815
ISBN-10: 1519370814

BOOKS BY TAYYIP ORAL

1. T. Oral and Dr. S. Warner, 555 Math IQ Questions for Middle School Students, USA, 2015

2. T. Oral, Dr. S. Warner, Serife Oral, Algebra Handbook for Gifted Middle School Students, USA, 2015

3. T. Oral, Geometry Formula Handbook, USA, 2015

4. T. Oral and Sevket Oral, 555 Math IQ Questions for Elementary School Students, USA, 2015

5. T. Oral, IQ Intelligence Questions for Middle and High School Students, USA, 2014

6. T. Oral, Dr. S. Warner, Serife Oral, 555 Geometry Problems for High School Students, USA, 2015

7. T. Oral, 555 ACT Math Problems, USA, 2015

8. T. Oral, E. Seyidzade, Araz publishing, Master's Degree Program Preparation (IQ), Cag Ogretim, Araz Courses, Baku, Azerbaijan, 2010, Azerbaijan.

9. T. oral, M. Aranli, F. Sadigov, and N. Resullu, A Text Book for Job Placement Exam in Azerbaijan for Undergraduate and Post Undergraduate Students in Azerbaijan, Resullu publishing, Baku, Azerbaijan - 2012 (3.edition)

10. T. Oral and I. Hesenov, Algebra (Text Book), Nurlar Printing and Publishing, Baku, Azerbaijan, 2001.

11. T. Oral, I. Hesenov, S. Maharramov, and J. Mikaylov, Geometry (Text Book), Nurlar Printing and Publishing, Baku, Azerbaijan, 2002.

12. T. Oral, I. Hesenov, and S. Maharramov, Geometry Formulas (Text Book), Araz courses, Baku, Azerbaijan, 2003.

13. T. Oral, I. Hesenov, and S. Maharramov, Algebra Formulas (Text Book), Araz courses, Baku, Azerbaijan, 2000

555 Advanced Math Problems

for Middle School Students

450 Algebra Questions and
105 Geometry Questions

Dr. Steve Warner, Tayyip Oral

Table of Contents

555 Math Questions   7

Test 1   7
Test 2   12
Test 3   17
Test 4   22
Test 5   27
Test 6   32
Test 7   37
Test 8   42
Test 9   47
Test 10   52
Test 11   57
Test 12   62
Test 13   67
Test 14   72
Test 15   77
Test 16   82
Test 17   87
Test 18   92
Test 19   97
Test 20   102
Test 21   107
Test 22   112
Test 23   117
Test 24   122
Test 25   127
Test 26   132
Test 27   137
Test 28   142
Test 29   147
Test 30   152

Test 31   157

Test 32   162

Test 33   167

Test 34   172

Test 35   177

Test 36   182

Test 37   187

Answer Key   **192**

About the Author   ***196***

Books by Tayyip Oral   197

Books by Dr. Steve Warner   198

555 MATH QUESTIONS

Test 1

1. $0.\overline{12} + 0.\overline{2} + 0.\overline{32} =$

2. A basketball team lost $\frac{1}{3}$ of its matches and tied $\frac{1}{4}$ of its matches. If the team won 30 matches, how many matches did the team tie?

3. Jack purchased a new ACT math book at a 20% discount for $120. What was the original price of the book?

4. The sum of two consecutive integers is 41. Find the product of these integers.

5. A school has a total of 120 students, 40 of whom are girls. What is the ratio of girls to boys in the school?

6. If there is a 20% discount on an item whose selling price is $160, what is the dollar amount of the discount?

7. If 38 liters of water is poured into 0.6 liter bottles, how many bottles would be filled completely?

8. On an exam, thirty students scored 80 points and forty students scored 90 points. Find the mean score of the seventy students.

9. There are thirty houses on a street containing a total of 110 rooms. If each house has 3 or 4 rooms, how many of the houses have 4 rooms?

10. There are 48 men and 12 women in a room. How many married couples would need to enter the room so that there will be three times as many men as women?

11. When the number 5.375 is written in the form $a\frac{b}{c}$ with GCD $(b, c) = 1$, what is the value of $a + b + c$?

$$a + b = 16$$
$$a + c = 14$$
$$b + c = 12$$

12. Find the unique solution (a, b, c) to the system of equations displayed above.

13. 20 kilograms of olive oil can be made from 24 kilograms of olives. How many kilograms of olive oil can be made from 96 kilograms of olives?

	Spanish	German	French	Total
Male	70	10	12	92
Female	110	30	3	143
Total	180	40	15	235

14. The table above shows the first language of 235 students classified by gender. What is the probability that a randomly selected French student will be female?

15. Let x and y be nonzero real numbers. If $\frac{7x+4y}{x} = 8$, then what is the value of $\frac{4y+x}{y}$?

Test 2

1. The number of people who visited a museum increased from 24,000 in 2011 to 28,000 in 2012. Find the percent increase.

2. The sum of two numbers is 16 and their difference is 2. Find the smaller number.

3. If a and b are positive integers, what is the maximum possible value for b if $5a + 7b = 700$?

4. Suppose that $-6 < a < 0 < b < 8$ where a and b are real numbers. Describe all possible real values of $b - a$.

5. Let a and b be positive integers satisfying $\frac{a}{5} + 2b = 11$. What is the maximum possible value of a?

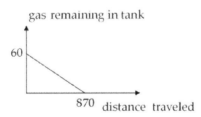

6. The graph displayed above shows the gas remaining in a vehicle's tank, in liters, while travelling. At what distance will the vehicle have 40 liters in the tank?

7. If $\frac{a}{b} = 4.25$ and GCD(a,b)= 1, then $(a + b)^2 =$

8. If $7^n = x$, then find 7^{n-2} in terms of x.

9. If $7^n + 5^n = a$ and $7^n - 5^n = b$, compute $7^{2n} - 5^{2n}$.

10. If $3^n \cdot 5^n = n$, find $9^n \cdot 25^n$.

11. Let a, b, and c be positive integers with $\frac{a}{3} = \frac{b}{7} = \frac{c}{11}$. What is the smallest possible 2-digit value of $a + b + c$?

12. If $3x - 3y = 12$, then $2x - 2y =$

13. Two vehicles are travelling at speeds of 60 km/hr and 90 km/hr, repectively, from location A. If the slower vehicle arrives at location B four hours later than the faster vehicle, then what is the distance between locations A and B?

14. If a bookstore sells its books for $.30 each it will make a $4.30 profit, and if the bookstore sells its books for $.18 each it will take a $.50 loss. How many books does the bookstore have?

15. The sum of two numbers is 32 and their difference is 2. What is the value of the smaller number?

Test 3

$$\sqrt{5}, \sqrt{20}, \sqrt{45}, \sqrt{80}$$

1. Find the arithmetic mean of the four numbers listed above.

2. In a large university stadium the ratio of men to women is 4:5. If there are 3,600 people in the stadium, find the positive difference between the number of women and men.

3. Find the sum of all possible three digit numbers that contain each of the digits 3, 4, and 5 exactly once.

4. The positive difference between the squares of two consecutive odd integers is 24. Find the value of these integers.

5. What is the smallest positive integer k such that 48 and 50 are both divisible by k ?

$$\frac{16}{17} \quad \frac{17}{18} \quad \frac{27}{28} \quad \frac{9}{10} \quad \frac{36}{37}$$

6. Which of the fractions shown above is greatest?

7. Let x be a positive real number such that $x + \frac{9}{11}$ is a positive integer. What is the least possible value for x ?

$$2a + 3b = 12$$
$$3a + 2b = 13$$

8. In the system of equations above, $a + b =$

9. If $\frac{a}{5} + \frac{b}{7} + \frac{c}{11} = 15$, then $77a + 55b + 35c =$

10. An integer between 200 and 260 has 14 and 21 as factors. What is the minimum possible value of the sum of this integer's digits.

11. If $\frac{a}{b} = \frac{c}{d} = \frac{e}{f} = 5$, then $\frac{a+b}{b} \cdot \frac{c+d}{d} \cdot \frac{e+f}{f} =$

12. Three brothers have ages in the ratio of 2:4:6. If the sum of the brothers' ages is 60, find the eldest brother's age.

13. If a car travels from location A to location B at 100 km/h the driver will arrive 15 minutes late. If the car travels 120 km/h the driver will arrive 40 minutes early. What is the distance between locations A and B?

$$14, 13, 26, 25, 50, x, y$$

14. In the sequence shown above, what is the value of $x + y$?

15. $0.4 + 0.08 + 0.006 =$

Test 4

Country	Numbers of matches	Number of fans
Russia	4	9000
Ukraine	6	15000
Belgium	8	28000
Italy	14	26000
Turkey	17	90000

1. The table above shows the countries that were represented in a sports competition, the number of matches played by the team from each country, and the number of fans that attended those matches. Which country had the most fans per match?

2. A bucket is filled halfway with water. Water is removed with an 800 ml container eight times, and then one sixth of the bucket is filled with water. How many liters of water can the bucket hold?

3. The sum of the ages of 6 siblings is 80. What was the sum of their ages 4 years ago?

4. Jack can complete a given task in 12 days. George can complete the same task in 16 days. If the two men work together, what fraction of the task can they complete in 1 day?

Semester	2009	2010
September	500	300
October	600	420
November	800	820
December	900	920
January	1100	1220
Febuary	700	770

5. The table above shows the number of students that took a specific exam during a semester lasting from September through February in the years 2009 and 2010. What was the percentage increase in the number of students from February 2009 to February 2010?

6. Let x and y be positive integers, and let

$$A = \frac{x+1}{y}, B = \frac{x+3}{y}, \text{ and } C = \frac{x+1}{y+4}.$$

List A, B, and C in increasing order.

7. Let a and b be positive integers satisfying $a \cdot b = 21$ and $c = 7b$. What is the value of $a + b + c$ when b is as large as possible?

8. One hose can fill an empty pool in 12 hours, and a second hose can fill the same pool in 20 hours. If both hoses are used together, in how many hours can the pool be filled?

9. One package of bird feed will last a canary 100 days, a pigeon 60 days, and a chicken 40 days. How many days will the package of feed last if it is used to feed all 3 birds?

10. 10 girls and 20 boys took an exam. The girls scored an average of 106 points and the boys scored an average of 98 points. What was the overall average score of the girls and boys together?

11. Find the sum of all possible 3 digit numbers that contain the digits 2, 3, and 4.

12. There are 64 men and 16 women in a room. How many married couples would need to enter the room so that there will be three times as many men as women?

13. A class has a total of 44 students of which 24 are girls. 12 of the girls and 13 of the boys have blonde hair. What is the probability that a randomly chosen student will be a blonde girl or a boy that is not blonde?

14. A 22 liter sugar water solution that contains 12% sugar and an 82 liter sugar water solution that contains 22% sugar are mixed together. What is the percentage of sugar in the new solution to the nearest tenth of a percent?

15. A taxi stops at a corner every 24 minutes and a bus stops at the same corner every 60 minutes. If the bus and taxi are both at the corner at 12 PM, then what is the next time when they will meet?

Test 5

1. At an airport an airplane takes off every 40 minutes from gate 1 and an airplane takes off every 60 minutes from gate 2. If airplanes take off from both gates simultaneously at 7 AM, when is the next time that airplanes will take off from both gates?

2. $875.889 - 874.888 =$

3. If $\frac{3x+7}{5} = \frac{3y+8}{4}$, then $12x - 15y =$

4. There are 4 yellow marbles, 3 red marbles, 5 green marbles, and 4 blue marbles in a box. What is the probability that a randomly chosen marble will be green?

5. There are 4 yellow marbles, 3 red marbles, 5 green marbles, and 4 blue marbles in a box. A marble is taken out of the box, and then without replacing the first marble another marble is taken out. What is the probability that the first marble is yellow and the second marble is red?

6. What is the ratio of $6 + \frac{1}{6}$ to $7 - \frac{1}{7}$?

7. $\dfrac{0.011+0.022}{0.011} =$

8. If $\sqrt{50} + \sqrt{75} = a(\sqrt{b} + \sqrt{c})$ where b and c are prime, then $a + b + c =$

9. If $\dfrac{8}{3} - 5x = \dfrac{1}{2} + 5y$, then $x + y =$

10. If $4x - 9y + 30 = 0$ and $8x - 9y + 30 = 20$, then $x =$

11. A plant's height is measured every year. Each year the height of the tree is 20 cm more than the height of the plant the previous year. If the plant's height was 180 cm during it's 6th year, then what was the plant's height in its 4th year?

12. If $2x + 3y = 24$ and $3x + 2y = 21$, then $x + y =$

13. Simplify $-(-6) - 6(\pi - 6) - 36 - 6\pi$.

14. Two vehicles are travelling around a circular track with circumference 140 m. The vehicles begin at the same location and travel in opposite directions at speeds of 8 m/s and 6 m/s, respectively. In how many seconds will the vehicles collide?

Item	Crop land (%)	The amount of crops (%)
Tomatoes	50	60
Carrots	15	40
Lettuce	16	15
Potatoes	19	20

15. The table above shows the percentage distribution of 1600 tons of crops on 3000 acres of land. How many acres of land are being used for the tomatoes?

Test 6

1. Apples are removed from a bag and split into groups of 4, 6, and 10. If there are 2 apples remaining in the bag, what is the least number of apples that could have initially been in the bag?

2. A gumball machine has 7 red gumballs, 6 yellow gumballs, and 3 blue gumballs. If two gumballs are selected from the machine at random, what is the probability that they are both yellow?

3. Let A and B be positive integers such that thirty times the larger number is equal to seventy times the smaller number. What is the minimum possible value of $A + B$?

4. Let $0 < x < 1$, $a = \frac{1}{3x}$, $b = \frac{x^2}{3}$, and $c = 4x$. List a, b, and c in increasing order.

5. If 600 grams of dried apricots cost \$3.60, then what is the cost, in dollars, of 1200 grams of dried apricots?

6. A building contains 30 apartments with a total of 102 rooms. If each apartment has 3 or 4 rooms, how many of the apartments have 3 rooms?

7. A car travels v miles per hour from city A to city B and then without stopping travels $\frac{v}{2}$ miles per hour back. If the total trip takes 16 hours, how long does it take the car to go from city A to city B ?

8. If $3 < x < 6$ and $4 < y < 6$, then what are all possible values of $x - y$?

9. How many times greater is 50 than 0.4 ?

10. A clock is running slow and falls behind 10 minutes every 2 hours. If the clock has the correct time on Tuesday at 8 AM, what time will the clock read on Thursday at 8 AM?

11. $\dfrac{0.55}{0.011} + \dfrac{0.66}{0.06} =$

12. How many even integers are there between 17 and 113?

13. How many odd integers are there between 18 and 128?

14. There are 4 yellow, 7 red, 5 green, and 4 blue marbles in a box. If one marble is randomly chosen, what is the probability it will be green?

15. Joseph has 400 more dollars than Alex. If Joseph spends 200 dollars, he will have three times as much money as Alex. How much money, in dollars, does Alex have?

Test 7

1. A sports league has 12 teams. Every team plays a game against each other team exactly once. What is the total number of games played?

2. $(-3)^0 + (-3)^1 + (-3)^2 + (-3)^3 =$

3. A bookstore sells $\frac{3}{4}$ of its books for $5 and the remaining books for $10. If the bookstore sells $400 worth of books, what is the total number of books sold?

4. There are an equal number of boys and girls in a room. 35% of the boys are Italian and 45% of the girls are Italian. If there are 160 Italians, how many people are in the room?

5. $333 + 33.3 + 3.33 + 0.333 =$

6. The sum of two natural numbers is 22 and their difference is 6. What is the ratio of the smaller number to the larger number?

$$4x - 6y + 8z = 14$$
$$6x - 4y + 2z = 6$$

7. If the above equations are true, then $x - y + z =$

8. If $10x = 6y$ and $4y = 2z$, then find the ratio $x : y : z$.

packet	1	2	3	4	5
weight	110	120	122	121	123

9. The table above shows the weight, in grams, of 5 packets of walnuts. What is the average weight of the 5 packets?

10. If $a: b = 7: 11$, evaluate $\frac{a+b}{b-a}$.

11. If $11x = 13y$, what is the ratio of x to y ?

12. How much more is $8a - 5b$ than $5b - 8a$?

13. $(-3)^2 - 3^2 + 3^0 + (-3)^2 + 3^1 + 24 =$

14. If the ratio of a to b is 4 to 13, then $\left(\frac{b-a}{b}\right)^2 =$

15. If $3(x + 4) = 2(x + 2)$, then $x =$

Test 8

1. $\left(\sqrt{7} + \sqrt{14} + \sqrt{21}\right) \div \left(1 + \sqrt{2} + \sqrt{3}\right) =$

2. $(8 - 4)^{\pi} + (12 - 8)^{\pi} + \left(\sqrt{64} - \sqrt{16}\right)^{\pi} =$

3. If $x + y = 12$ and $x = y$, then $3x + 2y =$

4. $1 + 4 + 7 + \cdots + 55 =$

$$6 + 12 + 18 + \cdots + 120$$

5. If each term of the above series is increased by 4, then how much is the sum increased by?

6. What is the perimeter of a rectangle with width 2^x and length 2^y ?

7. If $3a + 3b - 3c = 21$, then $5a + 5b - 5c =$

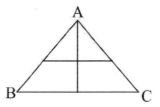

8. How many triangles are in the figure above?

9. Find half of $2^{2+\pi}$.

10. Clock A chimes every 10 minutes, clock B chimes every 15 minutes, and clock C chimes every 20 minutes. All 3 clocks chime together at 3 AM. What is the next time that all 3 clocks will chime together?

11. Three iron bars have lengths of 14 cm, 28 cm, and 35 cm. Each bar will be cut into pieces. What is the least total number of such pieces there can be so that all pieces have equal length?

12. The contents of a 50 kg bag of sugar, an 80 kg bag of salt, and a 120 kg bag of pepper will be distributed into bags without mixing the sugar, salt and pepper together. What is the least number of bags needed so that the contents of each bag have the same weight?

13. Find a number between $\frac{1}{7}$ and $\frac{3}{11}$.

14. If $a = 444$ and $b = 333$, then $\frac{(a-b)^2+4ab}{(a+b)^2-4ab} =$

15. If $a = 629$ and $b = 624$, then $(a + b)^2 - 4ab =$

Test 9

1. The arithmetic mean of three consecutive positive integers is 48. What is the least of the three integers?

2. One kilogram of oil costs $16. How much does 1800 grams of this oil cost?

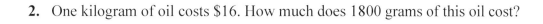

3. 8 more than 4 times a number is 88. What is the sum of this number's digits?

4. 6 times a number is 160 more than 2 times the number. What is the number?

5. The sum of two numbers is 96. One of the numbers is 5 times as large as the other one. What is the smaller of the two numbers?

6. The sum of three numbers is 162. The first number is 8 more than the second number. The second number is 4 less than the third number. What is the third number?

7. The cost of 7 notebooks and 3 pencils is $33. The cost of 4 notebooks and 8 pencils is $16. What is the cost of 1 notebook and 1 pencil to the nearest cent?

8. Each of the 30 rooms in a hotel has 2 or 3 beds. If there are a total of 78 beds in all the rooms in the hotel, how many rooms have 2 beds?

9. In a theatre, ticket prices are $6 for students and $8 for nonstudents. If 40 people buy tickets for a total of $290, how many of these people are students?

10. Several people are planning to go on a trip together. If each person were to pay $4000, then they would be $7000 short. If each person were to pay $5000, then they would be $8000 over. How many people are planning to go on the trip?

11. A man walks in a straight line. Every 7 steps he takes forward is followed by 3 steps backward. If he takes 270 steps altogether, how many steps forward did he take?

12. If $x + 2y = 3m$ and $x - 2y = m$, then find x in terms of m.

13. In 2003 a publishing house published 60,000 books. In 2011, the publishing house published 220,000 books. If the increase in the number of books each year was constant, how many books did this company publish in 2005?

14. A box filled with sand weighs 48 kg. The weight of the sand in the box is 7 times the weight of the box itself. What is the weight of the sand in the box, in kilograms?

15. When Jennifer spent $\frac{3}{7}$ of her money, she had $72 left. How much money did Jennifer start with?

Test 10

$$\frac{1}{3} \quad \frac{1}{2} \quad \frac{5}{9} \quad \frac{7}{12}$$

1. Find the arithmetic mean of the 4 fractions shown above.

2. $\frac{1}{8}$ of Tony's money is the same amount as $\frac{1}{9}$ of Joanne's money. If they have $680 in total, how much does Tony have?

3. A teacher spends $\frac{1}{6}$ of his salary on rent, $\frac{1}{5}$ of his salary on home improvement, and $\frac{1}{12}$ of his salary on utilities. If the teacher then has $3300 remaining each month, what is the monthly salary of this teacher?

4. If $\frac{2}{7}$ of a number is 88, what is $\frac{6}{7}$ of the same number?

5. If $\frac{x+4}{6} = \frac{10}{12}$, then $x =$

6. The ratio of Daniel's age 8 years ago to his age 8 years from now is $\frac{2}{3}$. How old is Daniel today?

7. The sum of the ages of 5 siblings is 75. What is the sum of their ages 6 years from now?

8. A mother is 40 years old and her daughter is 12 years old. In how many years will the sum of their 2 ages be 62?

9. Two siblings are 14 and 18 years old. In how many years will the sum of their ages be 10 times the difference between their ages?

10. Jeffrey can complete a job in 10 days. Tom can complete the same job in 14 days. If the two men work together, how many days will it take them to complete the job?

11. Christine and Daniella can finish a job together in 12 hours. If Christine can finish the same job herself in 18 hours, then how many hours would it take Daniella to finish the same job by herself?

12. Sandy takes 5 times as long as Grace to complete a certain job. If the two of them working together can complete the job in 15 days, how many days would it take Sandy to complete the job working alone?

13. Ahmet works 4 times as fast as Veli. Working together they can finish a job in 20 days. How many days would it take for Veli to complete the job by himself?

14. Ali can complete a job in 8 days, Veli can do it in 10 days, and Hakan can do it in 15 days. If all three men work together, how long will it take them to complete the job?

15. Hose A can fill an empty pool in 7 hours and Hose B can fill the same empty pool in 14 hours. If the two hoses are used simultaneously, how long will it take them to fill the empty pool?

Test 11

1. Pipe A can fill an empty pool in 12 hours and Pipe B can fill the same empty pool in 18 hours. If the two pipes are used simultaneously, how long will it take them to fill the empty pool?

2. If 16 people can complete a job in 48 days, then how many days does it take for 24 people to complete the same job?

3. Ahmet can finish a job in x days and Mehmet can finish the same job in 16 days. If the two men work together, they can finish the job in $\frac{x}{4}$ days. What is the value x ?

4. Jonathon works 7 times as fast as Kenneth. If Jonathon and Kenneth can finish the job in 28 days working together, then how many days would it take for Jonathon to finish the job working alone?

5. If a man driving 60 km/h can get from Town A to Town B is 6 hours, then how long would it take him if he were driving 72 km/h?

6. A man drove 80 km/h from City A to City B, and then made the return trip from City B to City A at 60 km/h. If it took him 2 more hours to make the return trip, what is the distance between City A and City B?

7. A vehicle is travelling from Town A to Town B at 70 km/h, and another vehicle is travelling from Town B to Town A at 60 km/h. If the distance between the two cities is 390 km, then after how many hours will the two cars meet?

8. A vehicle travels between two cities in 8 hours. If the car's speed is decreased 25 km/h, then it travels the same distance in 10 hours. What is the distance between the two cities?

9. If $\frac{1}{a} + \frac{1}{b} + \frac{1}{c} = 3x$, then $\frac{3}{a} + \frac{3}{b} + \frac{3}{c} =$

10. 20% of 30% of what number is equal to 30?

11. 70% of the students in a certain class are male and 20% of the male students are on the football team. Under the assumption that none of the female students play football, what percentage of the students in the class are on the football team?

12. A $500 table was on sale for 20% off. A salesman then gave an additional discount of x%, selling the table for $200. What is the value of x ?

13. If a store were to sell all of its shirts for $20 per shirt, the store would make a $500 profit. If the store were to sell all of its shirts for $12 per shirt, the store would take a $220 loss. How many shirts are in the store?

14. What is 20% of 30% of 50?

15. The positive difference between 16% and 18% of a number is 40. What is the number?

Test 12

1. Find the sum of all 3 digit numbers that can be formed using each of the digits 1, 2 and 3 exactly once.

2. 72 grams of water are mixed with 32 grams of sugar. What is the sugar percentage of the solution?

3. In a class, 22 of the students know German, 15 of the students know Spanish, and 8 students know both languages. How many students know at least one of these two languages?

4. What is the sum of the even numbers between 1 and 61?

5. If $12 \leq x \leq 82$ and x is even, then how many possible values are there for x ?

6. x is a real number such that $x + \frac{4}{9}$ is a positive integer. What is the minimum possible value of x ?

7. If the radius of a circle is reduced by 60%, by what percentage is the area of the circle reduced?

8. If $a:b = 2:5$ and $b:c = 4:2$, then $(5a + 4b):6c =$

$$A = \frac{322}{323} \quad B = \frac{232}{231} \quad C = \frac{21}{20} \quad D = \frac{15}{14}$$

9. Which of the fractions above is closest to 1?

10. A train travels 300 km in 6 hours. The train then decreases its speed by 10 km/h and travels the remaining distance in 4 hours. What is the total distance that the train travelled?

11. When two pipes are opened, an empty pool takes 8 hours to fill. Assuming that both pipes fill the pool at the same rate, how much time does it take to fill the pool when one of these pipes is opened?

$$a + b = 4$$
$$a + c = 2$$
$$b + c = 6$$

12. In the system of equations above, $a + b + c =$

13. $\left(\frac{0.02}{0.4} + \frac{0.04}{0.8} - \frac{0.08}{1.6}\right)^{-1} =$

14. If $x < 0$ and $|2x - 4| = 6$, then $x =$

$$|3x - 8| = 4$$

15. Solve the equation above for x.

Test 13

1. A father is 32 years old and his daughter is 8 years old. In how many years will the ratio of their ages be 5:2?

2. $6 + 8.1 - 48 \div 12 =$

3. Write $\frac{18}{37}$ as a decimal.

4. Ahmet is 10 years old and Mehmet is 25 years old. In 5 years what will be the ratio of Ahmet's age to Mehmet's age?

5. $\frac{1}{2}+\frac{1}{3}+\frac{1}{4}+\frac{1}{5}=$

$$100, 121, 144, 169, x$$

6. In the sequence above, $x =$

$$7, 11, 13, 17, 19, x$$

7. In the sequence above, $x =$

$$1, 1, 2, 4, 3, 9, 4, 16, x, y$$

8. In the sequence above, $x + y =$

9. $2 + 4 + 6 + \cdots + 80 =$

10. Let x, y, and z be real numbers such that $(6x - 5y)^2 = 0$ and $(2y - 6z)^2 = 0$. Evaluate $\frac{6x+5y}{2y+6z}$

.

11. If the perimeter of a square is increased by 30%, by what percentage will the area of the square increase?

12. If $\frac{bc}{a} = 1$, $\frac{ca}{b} = 2$, and $\frac{ab}{c} = 3$, then $a^2 + b^2 + c^2 =$

13. How many digits does 2^{12} have?

14. A father is 51 years old and his daughter is 13 years old. In how many years will the father's age be 3 times his daughter's age?

15. Find the sum of the digits of the largest 3 digit positive integer that is divisible by 5, 6, and 8.

Test 14

1. If $\frac{a-b}{b} = 4$, then $\frac{a^3+ab^2}{a^2b+b^3} =$

2. $\frac{0.4}{0.08} - \frac{2.4}{0.48} + \frac{10.8}{0.18} =$

3. $\frac{0.6}{0.06} - \frac{0.8}{0.08} + \frac{0.9}{0.09} =$

4. $\frac{3}{4} + \frac{4}{3} + \frac{7}{5} - \left(\frac{3}{4} - \frac{4}{3} + \frac{7}{5}\right) =$

5. Solve for x: $|4x - 6| = 10$

6. Solve for x: $|3x - 4| \leq 6$

7. If $f(x) = 3x^2 + 3$, then $f(x + 2) =$

8. If $y = \frac{7x-3}{3x+4}$, what is x in terms of y?

9. When 10 kg of fresh apricots are dried, 8 kg of dried apricots are produced. How many kilograms of fresh apricots do we need in order to produce 24 kg of dried apricots?

$$x, 6, 14, 8, 10, y$$

10. If the arithmetic mean of the numbers in the above list is 10, then $x + y =$

11. Find the geometric mean of 6 and 12.

12. If Ahmet can complete a job in 20 days, and Mehmet can complete the same job is 30 days, how long would it take to complete the job if Ahmet and Mehmet were to work together?

13. $1^2 + 3^2 + 5^2 + \cdots 9^2 =$

14. Find the ratio of $\frac{1}{4}$ to $\frac{3}{5}$.

15. In a group of 66 people there are 7 times as many women as men after 12 men and 14 women leave the group. How many men were in the group initially?

Test 15

1. If $\frac{4}{11}$ of a number is 60, what is the number?

2. If $A = (x + y)^2 + 2xy + 3(x - y)$, then evaluate A when $x = 1$ and $y = 2$.

3.

$\frac{60}{40}$ $\frac{30}{70}$ $\frac{75}{25}$ $\frac{64}{x}$ **x=?**

4. $1.69 + 16.9 + 169 =$

5. How many positive divisors does 400 have?

6. If $\frac{a}{7} + \frac{b}{14} + \frac{c}{28} = 12$, then $4a + 2b + c =$

7. LCM(24,36) =

8.

9. The lengths of two adjacent sides of a rectangle are $x + 1$ and $x^2 + 1$. Find the area of the rectangle in terms of x.

$$7, 5, 8, 6, 9, 7, 10, x, y$$

10. In the sequence shown above, what is the value of $x + y$?

$$3, 6, 8, 16, 18, 36, 38, x, y$$

11. In the sequence shown above, what is the value of $x + y$?

12. If $\frac{a}{3} + \frac{b}{6} + \frac{c}{9} = 10$, then $6a + 3b + 2c =$

13. Dale, Rick, and Alfred's ages are in the ratio 2 to 3 to 5. If Dale is 5 years younger than Rick, then what is Alfred's age?

14. 70% of a product is sold, and then 40% of the remaining product is sold. What percentage of the original product has not been sold?

15. Ahmet can finish a job in 12 days, Mehmet can finish the same job in 16 days, and Ali can finish the job in 24 days. If all three work together, how many days will it take for them to finish the job?

Test 16

1. LCM(8, 10, 15) =

2. GCD(8, 12, 36) =

3. If the longer side of a rectangle is increased by 60% and the shorter side of the rectangle is decreased by 80%, by what percentage is the rectangle's area decreased?

4. If $2 \leq m \leq 5$ and $2 \leq n \leq 3$, then what is the maximum value of $\frac{3}{m} + \frac{n}{2}$?

5. Emir is 4 years older than Erhan and 4 years younger than Nazım. If Nazım's age is 4 less than 2 times Erhan's age, then how old is Emir?

6. Pump A is twice as powerful as Pump B. When the two pumps are activated, an empty pool takes 12 hours to fill. How many hours would it take to fill the pool if only Pump B were used?

7. If $6x + 4y = 20$ and $2x + 2y = 4$, then $xy =$

8. $\frac{7}{3} + \frac{6}{5} + \frac{8}{9} - \left(\frac{8}{9} + \frac{6}{5} - \frac{7}{3}\right) =$

9. Find the solution set of following inequality: $|3x - 7| \leq 14$

10. If $f(x) = x^2 + 3x$, then $f(x + 1) - f(x - 1) =$

11. If $x + y = 14$ and $x - y = 2$, then $2x + 3y =$

12. If $81^{2x+4} = 27$, then $x =$

13. $4^2 + 4^1 + 4^0 =$

14. If seven years ago the arithmetic mean of 6 brothers' ages was 10, then what is the arithmetic mean of their ages now?

15. When two pumps are activated, an empty pool takes 12 hours to fill. One of the pumps works 3 times faster than the other. How many hours would it take to fill the pool if only the faster pump were used?

Test 17

1. $0.\overline{2} + 0.\overline{3} + 0.\overline{5} =$

2. $\dfrac{0.\overline{1} + 0.\overline{2}}{0.\overline{3} + 0.\overline{4}} =$

3. $\text{GCD}(24, 32) =$

4. What is the sum of the prime numbers between 30 and 40?

5. $4 \cdot 3 + 6 - 5^2 + 12 \div 3 - (-6) =$

6. If $A = \{1, 2, 3, 4, 5, 6\}$ and $B = \{5, 6, 7, 8, 9\}$, then find $A \cap B$.

7. $\dfrac{3!+4!}{4!-3!} =$

8. If $121^x = m$, then $11^{3x} =$

9. $(10^5 + 10^4) \div (10^4 + 10^3) =$

10. $1^2 + 2^2 + 3^2 + \cdots + 10^2 =$

11. An item is discounted 25% during the first week of a sale. The item is discounted an additional 20% during the second week. By what percentage has the item been discounted from its original selling price?

12. Orhan can complete a job by himself in 20 days. Orhan and Ayhan can complete the job together in 12 days. How long would it take for Ayhan to complete the job by himself?

13. A hose can fill an empty pool in 24 hours, and a second hose can fill the pool in 8 hours. If both hoses are used simultaneously, how long will it take to fill the pool?

$$a = \frac{20}{72}, \quad b = \frac{32}{36}, \quad c = \frac{35}{44}$$

14. Given the fractions above, arrange a, b, and c in increasing order.

15. Let a, b, and c be positive real numbers with $a + b = \frac{6}{5}$, $b + c = \frac{7}{5}$, and $a + c = \frac{4}{3}$. Arrange a, b, and c in increasing order.

Test 18

1. $(6^x + 6^x + 6^x + 6^x) \div (3^x + 3^x + 3^x + 3^x) =$

2. Let f be a function such that $7^{f(x)} = x + 5$. Evaluate $f(2) + f(44)$.

3. $(4^2 + 4^1 + 4^0) \div (3^2 + 3^1 + 3^0) =$

4. If $\frac{a+b}{a-b} = 9$, then $\frac{a}{b} =$

5. $(2!)^2 + (3!)^2 =$

6. Find the arithmetic mean of $4\sqrt{7} - 3$ and $4\sqrt{7} + 3$.

7. A hose can fill an empty pool in 8 hours, a second house can fill the same pool in 10 hours, and a third house can fill the pool in 12 hours. If all 3 hoses are used simultaneously, how long does it take to fill the pool?

8. The product of 2 prime numbers is 33. What is the sum of the two prime numbers?

9. GCD(13, 17) =

10. In how many ways can 3 books be chosen from 8?

11. If $x + y = 7$ and $x - y = 1$, then $x^2 + y^2 =$

12. $14 - \frac{0.16}{0.016} + \frac{0.24}{0.024} + \frac{0.14}{0.007} =$

13. If 200 grams of water is added to 500 grams of a 35% salt solution, then what is the percentage of salt in the new solution?

14. If $3mx - 2n = 4nx - 6$, then $x =$

15. If $x = m + n$, $y = m - n$, and $z = m^2 - n^2$, then $\left(\frac{x+y-2n}{z}\right) =$

Test 19

1. LCM(24,72) =

$$A = \{1, 3, 5, 7, 8, 9\} \quad B = \{2, 4, 6\}$$

2. Given sets A and B above, $A \cup B =$

3. $144 + 14.4 + 1.44 + 0.144 =$

4. $0.\overline{4} + 0.\overline{6} + 0,\overline{7} =$

5. $|-8 - 10| + |-4 - 7| =$

6. $|-2.44 - 2.56| + |-1.24 - 1.76| =$

k	34	62	81	99	66
$f(k)$	7	8	9	18	x

7. The table above defines a function f. What is the value of x?

k	92	84	61	72	93
$f(k)$	7	4	5	5	x

8. The table above defines a function f. What is the value of x?

$$4^3 \quad 4^{15} \quad 3^{15}$$

9. Arrange the numbers above in increasing order.

10. $\dfrac{4^{15}+4^{15}+4^{15}+4^{15}}{2^{15}+2^{15}+2^{15}+2^{15}} =$

11. Jennifer bought a textbook with 2/7 of her money. She then bought a notebook with 3/5 of her remaining money and was left with $12. How much money did she start with?

12. A train that is 80 meters long travels 240 meters in 6 seconds. How many seconds would it take the train to pass through a 480 meter long tunnel?

13. If $14x + 15 = 0$, then $14|x| =$

14. If $\dfrac{1}{x} + \dfrac{1}{2x} + \dfrac{1}{3x} = \dfrac{11}{30}$, then $x =$

15. If $(2x + 6)^{10} = (x + 3)^{10}$, then $x =$

Test 20

1. The ratio of men to women in a movie theatre that is filled to capacity is 8:7. If the maximum capacity of the theatre is 270 people, how many women are in the theatre?

2. LCM(40,90) =

3. GCF(60,80) =

4. $4^{-2} + 3^{-2} + 2^{-2} =$

5. If $324_{\text{five}} = x_{\text{four}}$, then $x =$

6. If $A = \{a, b, c, d, e, f\}$ and $B = \{a, b, k, l, m, n\}$, then $A \cap B =$

7. $7\frac{1}{2} \cdot 4\frac{1}{3} \cdot 6\frac{2}{3} =$

8. Five years ago the sum of two siblings' ages was 16. What is the sum of their ages today?

9. Find the geometric mean of 25 and 36.

$$\frac{2}{3}, \frac{3}{4}, \frac{5}{6}, \frac{7}{8}$$

10. Write the above fractions in increasing order.

$$
\begin{array}{c|ccccc}
k & 72 & 91 & 41 & 84 & 81 \\
\hline
f(k) & 25 & 64 & 9 & 16 & x
\end{array}
$$

11. The table above defines a function f. What is the value of x?

12. Toby is 160 cm tall and the length of his shadow is 480 cm. If the length of Phil's shadow is 510 cm, then what is Phil's height, in cm?

13. $(11^8 + 11^8 + 11^8 + 11^8 + 11^8) \div 121^4 =$

14. $\left(\sqrt{30} + \sqrt{40} + \sqrt{50} + \sqrt{60}\right) - \left(2\sqrt{10} + 2\sqrt{15}\right) =$

15. $3! + 4! + 5! =$

Test 21

1. If $\dfrac{a-2}{a-4} = \dfrac{a-6}{a-5}$, then $a =$

2. Two cars begin at the same location and travel in opposite directions at speeds of 80 km/h and 60 km/h, respectively. Find the distance between the two cars after 4 hours.

3. A father's age is 46 and his son's age is 10. In how many years will the father's age be 3 times the age of his son?

4. If $2x + y = 12$ and $3x - 2y = 11$, then $xy =$

5. Find all real values x such that $x + 3|x| - 6 = 0$.

$$\frac{7}{\sqrt{17} - \sqrt{10}}$$

6. Rewrite the fraction above so that no square roots appear in the denominator.

7. If f is a linear function such that $f(2) = 6$ and $f(3) = 8$, then $f(10) =$

8. The arithmetic mean of a student's 3 exam grades is 84. What does the student need to score on his fourth exam to bring his mean score up to 88?

9. What is the product of all possible solutions to the equation $(x + 4)^2 = 25$?

10. Find a fraction between $\frac{1}{3}$ and $\frac{2}{7}$.

11. If $|2x + 3| = 6$, then $x =$

12. $|6 - 3\pi| + |3\pi - 6| =$

13. If $\sqrt{6} + \sqrt{15} = x\sqrt{15}$, then $x =$

14. If $10a - 6b + 2c = 0$ and $8a + 2b - 4c = 0$, then $a:b =$

15. If $a + b = 9$ and $a^2 + b^2 = 41$, then $ab =$

Test 22

1. $1^3 + 2^3 + 3^3 + \cdots + 10^3 =$

$$\frac{29}{a} < \frac{11}{7}$$

2. What is the smallest natural number a that satisfies the above inequality?

3. 3 kg of salt is added to 15 kg of pure water. What percentage of the mixture is salt?

4. The price of a product was increased by 10% in January, and then by 20% in February. In March the product's price was decreased by 25%. By what percentage has the initial price changed?

5. If $\frac{10!-10\cdot 8!}{8!} = 2^x 5^y$, then $x + y =$

6. A father's age is 4 times his son's age. In 5 years the sum of their ages will be 60. How old is the son?

7. If $3x + y = 7$ and $x + 3y = 19$, then $(x + y)^2 =$

8. If $\frac{6}{2x} - \frac{2}{2y} = 4$ and $\frac{4}{x} + \frac{6}{y} = 8$, then $\frac{y}{x} =$

9. If $0 < x < 1$, $a = \frac{x}{2}$, $b = x^3$, and $c = \frac{2}{\sqrt{x}}$, then list a, b, and c in increasing order.

10. If $x < 0$, then rewrite the expression $|x - 2| + |2x| + 6$ without using absolute values.

11. An employee works 6 hours per day. If we use a circle graph to display this information, then how many degrees will the central angle measure be that corresponds to the number of hours that the employee works per day.

12. What percentage of 300 is 15?

13. If $25^{x+2} = 1$, then $x =$

14. The sum of two numbers is 26 and the difference of their squares is 52. What is the smaller of the two numbers?

15. What is half of 40% more than 400?

Test 23

1. If $x < 0$, then $\dfrac{|-4x|}{x} + \dfrac{|6x|}{x} =$

2. If $314_{\text{five}} - 222_{\text{five}} = x_{\text{five}}$, then $x =$

3. $(10 + 12.3 - 30 \div 3) \div (15 + 2.6 - 9 \div 3) =$

4. $\dfrac{13^{83}-13^{82}}{13^{84}} =$

5. An exterior angle of a regular polygon measures 20°. How many sides does the polygon have?

6. If $a > 0$, then rewrite $|6a| + |-4a|$ without absolute values.

7. If $\frac{2}{7} = \frac{x}{10}$, then $x =$

8. If $\frac{x+1}{5} = \frac{6}{10}$, then $x =$

9. LCM(3,4,7)=

10. $\dfrac{3+\frac{1}{2}}{3-\frac{1}{2}} =$

11. $\left(4 + \frac{1}{2}\right) - \left(4 - \frac{1}{2}\right) =$

12. $42.24 - 2.2 =$

13. $2.12 + 3.42 + 6.36 =$

14. $2\frac{1}{3} + 3\frac{1}{4} + 4\frac{1}{5} =$

15. If $a - \frac{1}{a} = m$, then $a^2 + \frac{1}{a^2} =$

Test 24

1. $1 + 3 + 5 + \cdots + 23 =$

2. \$650 will be shared among Ayşe, Fatma, and Nurgül according to the following two rules:

  - Ayşe will get \$150 more than Fatma
  - Fatma will get twice as much as Nurgül

How much money will Fatma get?

3. If we decrease a number by 15% after we have just increased the number by 15%, then by what percentage is the original number decreased?

4. If $a:b = 3:4$ and $4a + 3b = 84$, then $a^2 + b^2 =$

5. Three years ago, the ages of three siblings added up to 24. What will be the sum of their ages in 6 years?

6. 15 kg of salt is mixed with 25 kg of water. What is the percentage of salt in the mixture?

7. $0.\overline{24} + 0.\overline{42} =$

8. If $433_{\text{five}} - 222_{\text{three}} = x_{\text{five}}$, then $x =$

9. $(4^{-2} + 4^0)^{-1} + (3^{-2} + 3^0)^{-1} + (2^{-2} + 2^0)^{-1} =$

10. If $x + \dfrac{6}{y} = 14$ and $x - \dfrac{6}{y} = 2$, then $y =$

11. What is the ratio of the number of weeks in a year to the number of days in a year (assume the year is not a leap year)?

12. If $(4x + 5)^8 = 1$, then $x =$

13. LCM(10,12,18) =

14. $(2! + 3! + 4!) \div (3! + 4! + 1!) =$

15. How many positive integers between 1 and 100 have 7 as a digit?

Test 25

1. A tree's height increases by 25% each year. In 4 years the height of the tree will be 625 cm. What is the height of the tree now?

2. $(14! + 13! + 12!) \div 12! =$

3. $123_{\text{five}} = x_{\text{ten}}$. Find x.

4. $11_{two} \cdot 22_{three} \cdot 33_{four} = x_{ten}$. Find x.

5. $(9 \cdot 10^9) \div (3 \cdot 10^3) =$

6. $(120 \cdot 10^7) \div (24 \cdot 10^2) =$

7. $\left(\sqrt{22} + \sqrt{88}\right) \div \sqrt{22} =$

8. $\left(\sqrt{12} + \sqrt{75} + \sqrt{27}\right) \div \sqrt{3} =$

9. If $3x - 9 = 21$, then $x =$

10. If $3(x - 2) = 2(x + 4)$, then $x =$

11. If $4x + y = x + 3y$, then $3x - 2y =$

12. If $7^x = m$, then $343^x + 7^{x+2} =$

13. Find $\frac{1}{4}$ of 8^8.

14. If $3x + 4y = 17$ and $x + 4y = 16$, then $xy =$

15. What is the minimum real value of $|2x + 8| + |4x - 10|$?

Test 26

1. If $\frac{x}{4} = \frac{6}{8}$, then $x =$

2. $(a^7 b^8 c^6) \div (a^2 b^4 c^2) =$

3. $243\Delta \rightarrow 8$, $369\Delta \rightarrow 4$, $246\Delta \rightarrow 4$, $993\Delta \rightarrow ?$

4. $123\Delta \rightarrow 6$, $124\Delta \rightarrow 8$, $246\Delta \rightarrow 48$, $443\Delta \rightarrow ?$

5. If $3 : m = 4 : 44$, then $m =$

6. If $3x + 2y = 15$ and $2x + 3y = 25$, then $x + y =$

7. Simplify the expression: $(a^6 b^7)(a^3 b^3)$

8. Simplify the expression: $(4a^7 b^6 c^8)^2 =$

9. $1 + 3 + 5 + \cdots + 21 =$

10. If $a:b = c:d = e:f = k$, then $\left(\frac{a+2c+3e}{b+2d+3f}\right) =$

11. A man deposits $800 into a bank account. Each year he makes 20% of the amount he has in the account. After 3 years how much money will he have?

12. A hose can fill an empty pool in 10 hours and a second hose can fill the same pool in 12 hours. How many hours will it take to fill the pool if both hoses are used?

13. Let x, y, and z satisfy $(8x - 6y)^2 - (4y - 10z)^2 = 0$. Evaluate $\frac{8x+14y}{8y-10z}$.

14. If $18x + 16 = 0$, then $18|x| =$

15. If $\frac{a+3}{9} = \frac{10}{18}$, then $a =$

Test 27

1. A 400 gram solution contains 20% salt. How many grams of salt should we add to make the solution have 30% salt?

2. If 300 grams of salt is mixed with 400 grams of water, what is the percentage of salt in the mixture?

3. If $123_{\text{four}} + 321_{\text{four}} = x_{\text{five}}$, then $x =$

4. If $(6x + 7)^{10} = 1$, then $x =$

5. $0.007^3 \div 0.49^4 =$

6. $\sqrt{1.44} + \sqrt{1.21} + \sqrt{1.69} + \sqrt{2.25} =$

$$\sqrt{1}, \sqrt{9}, \sqrt{169}, \sqrt{400}, \sqrt{900}$$

7. Find the arithmetic mean of the 5 numbers shown above.

8. Factor $4a^2bc + 3ab^2c^2 + 10ab^2c^2$.

9. $(8x^2 + 4x) \div (4x) =$

10. $(9x^3 + 9x^2) \div (9x^2) =$

11. $(12ab^2 + ca^2b + 3ab) \div (3ab) =$

12. $(16a^2b^3) \div (12abc) =$

13. If $\frac{3x}{5} - 6 = 2x + \frac{8x-6}{10}$, then $x =$

14. The difference between two positive numbers is 2, and the product of these numbers is 120. What is the larger of the two numbers?

15. If $3553_{\text{six}} = x_{\text{ten}}$, then $x =$

Test 28

1. $\dfrac{6}{x+1} \cdot \dfrac{8x+8}{12} =$

2. $\dfrac{3x+4}{x} - \dfrac{2x+4}{x} =$

3. $\dfrac{x+5}{2y} + \dfrac{x-2}{3y} =$

$$\begin{array}{c|ccccccc} k & 6 & 7 & 9 & 12 & 16 & 21 & x \\ \hline f(k) & 3 & 4 & 6 & 9 & 13 & 18 & y \end{array}$$

4. The table above defines a function f. What is the value of $x - y$?

$$1, 8, 27, 64, x$$

5. In the sequence above what is the value of x ?

6. $1 + 3 + 5 + \cdots + 43 =$

7. $\frac{6}{11}$ of a tank is filled with water. After $\frac{1}{3}$ of the water in the tank is used, there will be 50 liters of water remaining. What is the tank's capacity in liters?

8. Let x and y be real numbers such that $-4 \le x \le 5$ and $-6 \le y \le 7$. Find the maximum possible value of $4x - 3y + 2$.

9. The sum of two numbers is 96. If 25% of one of the numbers is equal to 75% of the other number, what is the product of the two numbers?

10. Three friends have dinner at a restaurant. Orhan has $80 on him, Veli has $60, and Cemil has $40. If the three friends split the $126 check so that the amount they each pay is proportional to the amount they each have, then how much will Orhan pay?

11. The difference between a mother's age and her daughter's age is 40. If the mother is 5 times as old as the daughter, how old is the mother?

12. If $f(x) = 2x^2 + 3x + 40$, then compute the ratio of $f(2)$ to $f(3)$.

13. $0.\overline{12} + 0.\overline{22} + 0.\overline{32} =$

14. If $|a + 3| + |b + 4| + |c + 6| = 0$, then $2a + 3b + 2c =$

15. If $f(x) = x^2 + 3x + 4$, then $f(x + 2) =$

Test 29

$$\frac{2}{3}, \frac{3}{4}, \frac{1}{2}, \frac{5}{6}$$

1. Write the fractions above in increasing order.

2. $12\Delta \rightarrow 9, \quad 34\Delta \rightarrow 49, \quad 62\Delta \rightarrow 64, \quad 19\Delta \rightarrow 100, \quad 123\Delta \rightarrow ?$

k	123	369	444	524	755
$f(k)$	4	4	11	13	y

3. The table above defines a function f. What is the value of y?

4. $1 + 3 + 5 + \cdots + 45 =$

5. If $4x + \frac{1}{4x} = \frac{1}{3}$, then $16x^2 + \frac{1}{16x^2} =$

6. How many positive integer divisors does 77 have?

7. If $\frac{x-3y}{x+3y} = \frac{1}{3}$, then $\frac{x^3-4y^3}{x^3+4y^3} =$

8. Type A sugar sells for $3 per kg and Type B sugar sells for $2 per kg. 10 kgs of Type A sugar is mixed together with 15 kgs of Type B sugar, and the mixture sells for $4 per kg. How much more money is made from the mixture than if the same quantities of the two sugar types were sold separately.

9. Pipe A can fill an empty pool in 10 hours, Pipe B can fill the same empty pool in 12 hours, and Pipe C, at the bottom of the pool, can empty the pool in 24 hours. If the three pipes are used simultaneously, how many hours will it take to fill the empty pool?

10. If $64^{6n} = 16^5$, then $n =$

11. If $a = \frac{5}{14}$, $b = \frac{8}{27}$, and $c = \frac{13}{36}$, then list a, b, and c in increasing order.

12. If $a = \frac{7}{8}$, $b = \frac{5}{6}$, and $c = \frac{11}{12}$, then list a, b, and c in increasing order.

13. A solution is 40% salt. What is the ratio of water to salt in the solution?

14. The ratio of girls to boys in a class was 4:7. When 12 new girls were admitted to the class, the number of boys and girls became equal. How many students were in the class originally?

15. $((-3)^{300} \cdot 9^{100} \cdot 3^{-100}) \div 27^{50} =$

Test 30

1. If $f(x) = 2x + 3$ and $g(x) = x^2 + 4$, then $f(1) + g(1) =$

2. If $f(x) = 3x^2 + 2x$ and $g(x) = x^2 + 7$, then $\dfrac{f(2)}{g(2)} =$

3. If $x = 7^{10}$, $y = 49^6$ and $z = 343^6$ then list x, y, and z in increasing order.

4. Two saplings are planted next to each other. The first is 30 cm in height and the second is 20 cm in height. The first sapling grows 4 cm each year, and the second grows 2 cm each year. In 5 years, what is the difference between the heights of the two saplings?

5. Find the harmonic mean of 6 and 8.

6. Find the sum of the largest two-digit prime number's digits.

7. Find the product of the smallest and largest two-digit prime numbers.

8. The ratio of two prime numbers is 17:19. What is the least possible value of the sum of the two primes?

9. $(4^4 + 4^3 + 4^2 + 4^1) \div (2^4 + 2^3 + 2^2 + 2^1) =$

10. $\sqrt{1!} + \sqrt{2!} + \sqrt{3!} + \sqrt{4!} =$

11. If $f(x) = x^3 + 3x^2 + 3x + 3$, then $f(1) =$

12. $A \rightarrow 3, \quad L \rightarrow 2, \quad M \rightarrow 4, \quad N \rightarrow ?$

13. $123\Delta \rightarrow 6$, $142\Delta \rightarrow 8$, $331\Delta \rightarrow 9$, $444\Delta \rightarrow ?$

14. If $x > 4$, then $|x - 3| + |x - 2| + |2 - 2x| =$

15. $\sqrt{11 + 2\sqrt{10}} + \sqrt{11 - 2\sqrt{10}} =$

Test 31

1. Two angles are supplementary and the measure of one angle is twice the measure of the other. Find the measure of the smaller angle.

2. Two angles are supplementary and the measure of one angle is three times the measure of the other. Find the measure of the larger angle.

3. The ratio of the measures of supplementary angles is 4 to 6. Find the measure of the smaller angle.

4. Two angles are complementary and the measure of one angle is three times the measure of the other. Find the measure of the smaller angle.

5. The ratio of the measures of complementary angles is 2 to 3. Find the measures of the two angles.

6. The ratio of the measures of complementary angles is 1 to 4. Find the positive difference between the measures of the two angles.

7. The difference between the measures of complementary angles is 24. Find the measure of the smaller angle.

8. The difference between the measures of complementary angles is 36. Find the measure of the larger angle.

9. The ratio of the measures of complementary angles is 7 to 3. Find the positive difference between the measures of the two angles.

10. In an isosceles triangle, the measure of a base angle is 72°. Find the measure of the vertex angle.

11. The measures of the interior angles of a triangle are in the ratio 2:4:6. Find the measure of the smaller angle.

12. The angle measures of two interior angles of a triangle are 64° and 74°. Find the measure of the third interior angle.

13. The angle measures of two exterior angles of a triangle are 125° and 142°. Find the measure of the third exterior angle.

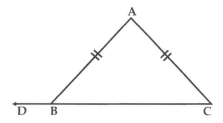

14. In triangle ABC above, $AB \cong AC$ and $m\angle ABD = 132°$. Find $m\angle A$.

15. In an isosceles triangle, the measure of the exterior angle adjacent to a base angle is 134°. Find the measure of the vertex angle.

Test 32

1. Find the length of a leg of an isosceles right triangle if the length of the hypotenuse is 11 cm.

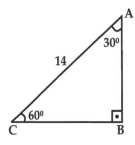

2. Find the perimeter of triangle *ABC* shown above.

3. Find the area of an equilateral triangle with side length 8 cm.

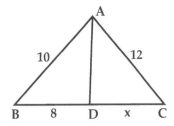

4. In the figure above *AD* is an angle bisector. Find x.

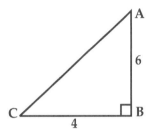

5. Find the area of right triangle *ABC* shown above.

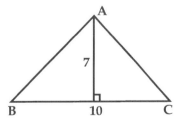

6. Find the area of triangle *ABC* shown above.

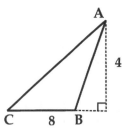

7. Find the area of triangle ABC shown above.

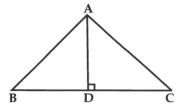

8. In the figure above, $BD = 12$ and $DC = 14$. Find the ratio of the area of triangle ABD to the area of triangle DAC.

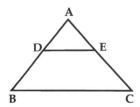

9. In the figure above, $DE \parallel BC$, $AD = 8$, $BD = 12$, $DE = 6$, and $BC = 2x + 6$. Find x.

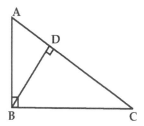

10. In the figure above, $m\angle B = m\angle D = 90°$, $m\angle C = 60°$, and $AB = 6$. Find the area of triangle ABC.

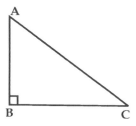

11. In the figure above, $m\angle B = 90°$, $m\angle C = 30°$, and $AC = 14$. Find AB.

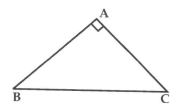

12. In the figure above, $m\angle A = 90°$, $m\angle B = 60°$, and $BC = 4$. Find AC.

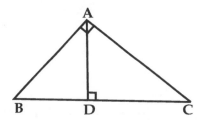

13. In the figure above, $m\angle A = m\angle ADC = 90°$, $m\angle C = 30°$, and $AB = 18$. Find AD.

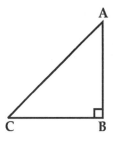

14. In the figure above, $AC = 61$ and $BC = 11$. Find AB.

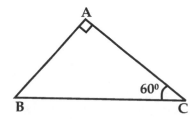

15. In the figure above, $m\angle A = 90°$, $m\angle C = 60°$, and $AC = 2$. Find AB.

Test 33

1. A triangle has sides of lengths 6 and 10. What is the maximum possible integer value for the length of the third side of the triangle?

2. In triangle ABC, angle bisector BD is drawn. If $AB = 18$, $AD = 8$, and $DC = 12$, then $BC =$

3. The ratio of the lengths of the sides of a triangle is 1:2:3. If the perimeter of the triangle is 28 cm, then what is the length of the shortest side of the triangle?

4. Find the area of an equilateral triangle with side length 11 cm.

5. Find the area of a square whose diagonal has length 12 cm.

6. Find the perimeter of a square whose diagonal has length 6 cm.

7. Find the area of a square whose perimeter is 7^x cm.

8. Find the area and perimeter of a square with side length 7^x cm.

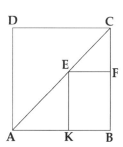

9. In square $ABCD$ shown above, $AB = 12$ cm. Find the perimeter of square $KBFE$.

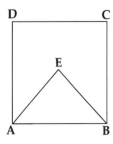

10. In the figure above, $ABCD$ is a square, ABE is an equilateral triangle, and $AB = 4$ cm. Find EB.

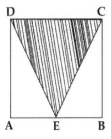

11. In the figure above, $ABCD$ is a square and the area of triangle DEC is 22 cm^2. Find the perimeter of $ABCD$.

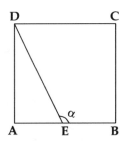

12. In the figure above, $ABCD$ is a square, $AE = 8$ cm, and $\alpha = 120°$. Find the area of $ABCD$.

13. The lengths of two sides of a rectangle are 8 cm and 10 cm. Find the perimeter, area, and length of a diagonal of the rectangle.

14. The perimeter of a rectangle is 80 cm. Find the area of the rectangle if the ratio of the lengths of adjacent sides is 1:3.

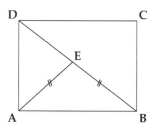

15. In the figure above, $ABCD$ is a rectangle, $AE = EB$, and $m\angle ADE = 56°$. Find $m\angle DEA$.

Test 34

1. Find the area of a rhombus with side length 10 cm and altitude 8 cm.

2. Find the area of a rhombus with diagonals of lengths 20 cm and 32 cm.

3. Find the area of a parallelogram with base 14 cm and height 6 cm.

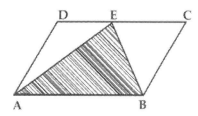

4. The area of parallelogram *ABCD* shown above is 33 cm^2. Find the area of triangle *AEB*.

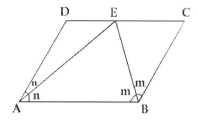

5. In parallelogram *ABCD* shown above, *AE* and *BE* are angle bisectors. If *AB* = 22 cm, then find the perimeter of *ABCD*.

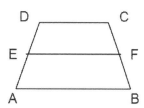

6. In trapezoid *ABCD* shown above, *E* is the midpoint of *AD*, *F* is the midpoint of *BC*, *AB* = 18, and *DC* = 12. Find *EF*.

173

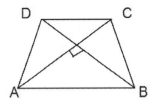

7. In trapezoid $ABCD$ shown above, $AB = 18$ cm and $DC = 12$ cm. Find the area of $ABCD$.

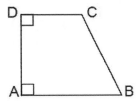

8. In trapezoid $ABCD$ shown above, $m\angle A = m\angle D = 90°$, $AB = BC = 18$ cm and $DC = 6$ cm. Find the area of $ABCD$.

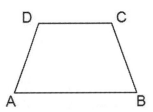

9. In trapezoid $ABCD$ shown above, $m\angle A = 50°$, $AB = 26$, $BC = 16$, and $DC = 10$. Find $m\angle C$.

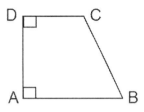

10. In trapezoid $ABCD$ shown above, $AB = BC$, $AD = 18$ cm, and $DC = 10$ cm. Find the area of $ABCD$.

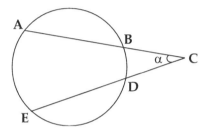

11. In the circle shown above, $m\widehat{AE} = 112°$ and $m\widehat{BD} = 44°$. Find α.

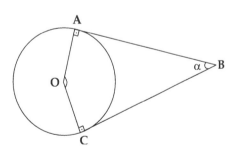

12. In the circle shown above with center O, $OA \perp AB$, $OC \perp BC$, and $m\angle AOC = 144°$. Find α.

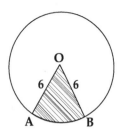

13. The circle above with center O has a radius of 6 cm and $m\angle AOB = 60°$. Find the area of the shaded sector.

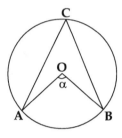

14. In the circle shown above with center O, $m\angle ACB = 44°$. Find α.

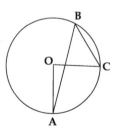

15. In the circle shown above with center O, $m\angle AOC = 76°$. Find $m\angle ABC$.

Test 35

1. Two of the exterior angles to a triangle measure 120° and 100°. What is the measure of the interior angle of the triangle that is nonadjacent to the given exterior angles?

2. The measures of the three angles of a triangle are in the ratio 2:3:5. What is the degree measure of the largest of the three angles?

3. Find the slope of the line that passes through the points $(3, -1)$ and $(1, 2)$.

4. The diagonal of a square is 6 cm long. Find the area of the square.

5. In triangle ABC, line segment BD bisects AC. If $AD = 2$ cm and $AB:BC = 8:3$, what is the length of CD ?

6. Two angles are supplementary and one is 25% as long as the other. Find the positive difference between the two angle measures.

7. In triangle ABC, AD bisects angle A, $m\angle BAC = 80°$ and $m\angle B = 40°$. Find $m\angle ADC$.

8. The perimeter of an equilateral triangle is 30 cm. Find the area of the triangle.

9. Two sides of a triangle have lengths of 5 cm and 8 cm. What lengths are possible for the third side of the triangle?

10. The diagonals of a rhombus have lengths 24 cm and 18 cm. Find the perimeter of the rhombus.

11. Two adjacent sides of a rectangle have lengths 2^x and 3^x. What is the area of the rectangle in terms of x ?

12. The hypotenuse of an isosceles right triangle is 10 cm long. Find the perimeter of the triangle.

13. Find the slope of the line with equation $y - 3x = 6$.

14. What is the slope of the line passing through the points (6,3) and (2,1)?

15. What is the distance from the origin to the point (5,-7)?

Test 36

1. A parallelogram has a base of length 7^x and a height of 5^x. Find the area of the parallelogram.

2. Find the radius of a circle that has area 400π.

3. Find the area of an equilateral triangle with side length 1.

4. Find an equation of the line that passes through the point (3,4) and has a slope of 2.

5. A right triangle has legs of length 5 cm and 12 cm. Find the ratio of the triangle's area to its perimeter.

6. The legs of an isosceles triangle each have length 34, and the height from the vertex angle to the base of the triangle is 30 cm. Find the length of the triangle's base.

7. An exterior angle of a regular polygon measures 60°. How many sides does the polygon have?

8. A rectangle has two adjacent sides of lengths 7^{2x} and 7^{3x}. What is the perimeter of the rectangle?

9. The diagonal of a square has length 6 cm. What is the ratio of the area of the square to the perimeter of the square?

10. A square has a side of length x cm and an area of 6 cm^2. What is the area of a square with side length $6x$ cm?

11. If the line with equation $y = 3x + a$ passes through the point (3,10), then $a =$

12. Let A be the sum of the measures of the interior angles of a triangle, and let B be the sum of the measures of the exterior angles of a triangle. Find the ratio of B to A.

13. Let A be the measure of an exterior angle of a regular hexagon, and let B be the measure of an interior angle of a square. Find the ratio of A to B.

14. Let A be the sum of the measures of the exterior angles of a square, and let B be the sum of the measures of the interior angles of a regular hexagon. What is the ratio of A to B?

15. Find the slope of the line that passes through the points $(-1,5)$ and $(4,1)$.

Test 37

1. A triangular pyramid has base area 12 cm^2 and the area of each of its other faces is 10 cm^2. Find the total surface area of the pyramid.

2. The length of the bottom edge of a square pyramid is 6 cm. What is the area of the base of the pyramid?

3. A right cone has a base radius of 8 cm and a height of 10 cm. Find the slant height of the cone.

4. Find the circumference of the base of a right cone that has a height of 8 cm and a slant height of 10 cm.

5. Find the lateral area of a right cone that has a height of 12 cm and a slant height of 13 cm.

6. Find the surface area of a sphere whose radius is 7 cm.

7. A rectangular prism has length 1 cm, width 3 cm, and height 4 cm. Find the surface area and volume of the prism.

8. Find the surface area and volume of a cube with edge length 6 cm.

9. Find the volume of a cube that has a surface area of 24 cm^2.

10. Find the surface area of a cube whose volume is 125 cm^3.

11. The edge length of two cubes is in the ratio of 2:3. Find the ratio of the volume of the smaller cube to the volume of the larger cube.

12. The edge length of two cubes is in the ratio of 7:5. Find the ratio of the surface area of the larger cube to the surface area of the smaller cube.

13. Find the volume of a triangular prism whose base area is $4\sqrt{3}$ cm^2 and whose height is $2\sqrt{3}$ cm.

14. Find the height of a triangular prism whose base area is 12 cm^2 and whose volume is 60 cm^3.

15. Find the surface area and volume of the solid generated by revolving a circle of radius 10 cm around its diameter 360°.

ANSWER KEY

Test - 1
1) $.\overline{6}$ or 2/3
2) 18
3) $150
4) 420
5) $\frac{1}{2}$
6) $32
7) 63
8) 600/7
9) 20
10) 6
11) 16
12) (9,7,5)
13) 80
14) 1/5
15) 8

Test - 2
1) 16 2/3%
2) 7
3) 95
4) $0 < b - a < 14$
5) 45
6) 290
7) 441
8) $x/49$
9) ab
10) n^2
11) 21
12) 8
13) 720 km
14) 40
15) 15

Test - 3
1) $5\sqrt{5}/2$
2) 400
3) 2664
4) 5, 7
5) 1200
6) 36/37
7) 2/11
8) 5
9) 5775
10) 3
11) 216
12) 30
13) 550 km
14) 147
15) 0.486

Test - 4
1) Turkey
2) 19.2
3) 56
4) 48/7
5) 10%
6) $C < A < B$
7) 169
8) 7.5
9) 19 11/31
10) 100 2/3
11) 1998
12) 8
13) 19/44
14) 19.9%
15) 2 PM

Test - 5
1) 9 AM
2) 1.001
3) 12
4) 5/16
5) 1/20
6) 259/288
7) 3
8) 10
9) 13/30
10) 5
11) 140 cm
12) 9
13) $6 - 12\pi$
14) 10
15) 1500

Test - 6
1) 62
2) 1/8
3) 10
4) $b < a < c$
5) $7.20
6) 18
7) 5 hrs 20 min
8) $-3 < x - y < 2$
9) 125
10) 6 AM
11) 61
12) 49
13) 55
14) 1/4
15) 100

Test - 7
1) 66
2) -20
3) 64
4) 400
5) 369.963
6) 4/7
7) 2
8) 3:5:10
9) 119.2
10) 9/2
11) 13/11
12) $16a - 10b$
13) 37
14) 81/169
15) -8

Test - 8
1) $\sqrt{7}$
2) $3 \cdot 4^\pi$
3) 30
4) 532
5) 80
6) $2^{x+1} + 2^{y+1}$
7) 35
8) 6
9) $2^{1+\pi}$
10) 4 AM
11) 11
12) 25
13) $\frac{12}{77}, \frac{13}{77}, \ldots, \frac{20}{77}$
14) 49
15) 25

Test - 9
1) 15
2) $28.80
3) 2
4) 40
5) 16
6) 54
7) $4.45
8) 12
9) 15
10) 15
11) 189
12) $2m$
13) 100,000
14) 42 kg
15) $126

Test - 10
1) 71/144
2) 320
3) 6000
4) 264
5) 1
6) 40
7) 105
8) 5
9) 4
10) 5 5/6
11) 36
12) 90
13) 100
14) 3 3/7
15) 4 hr 40 min

Test - 11
1) 1 hr 40 min
2) 32
3) 48
4) 32
5) 5 hours
6) 480 km
7) 3
8) 1000 km
9) $9x$
10) 500
11) 14%
12) 50
13) 90
14) 3
15) 2000

Test - 12
1) 1332
2) 30 10/13%
3) 29
4) 930
5) 36
6) 5/9
7) 84%
8) 2
9) A
10) 340 km
11) 16 hours
12) 6
13) 20/3
14) -1
15) {4/3, 4}

Test - 13
1) 8
2) 10.1
3) $.\overline{486}$
4) 1:2
5) 77/60
6) 196
7) 23
8) 30
9) 1640
10) 5/2
11) 69%
12) 11
13) 4
14) 6
15) 840

Test - 14
1) 65/13
2) 60
3) 10
4) 8/3
5) {-1,4}
6) $-\frac{2}{3} \le x \le 10/3$
7) $3x^2 + 12x + 15$
8) $\frac{4y+3}{7-3y}$
9) 30
10) 22
11) $6\sqrt{2}$
12) 12 days
13) 165
14) 5/12
15) 17

Test - 15
1) 165
2) 10
3) 36
4) 187.59
5) 15
6) 336
7) 72
8) 216
9) $x^3 + x^2 + x + 1$
10) 19
11) 154
12) 180
13) 25
14) 18%
15) 5 1/3

Test - 16
1) 120
2) 4
3) 68%
4) 3
5) 16
6) 36
7) -24
8) 14/3
9) $-\frac{7}{3} \le x \le 7$
10) $4x + 6$
11) 34
12) -13/8
13) 21
14) 17
15) 16

Test - 17
1) 10/9
2) 3/7
3) 8
4) 68
5) 3
6) {5,6}
7) 5/3
8) $m\sqrt{m}$
9)
10) 385
11) 40%
12) 30 days
13) 6 hours
14) $a < c < b$
15) $a < b < c$

Test - 18
1) 2^x
2) 3
3) 21/13
4) 5/4
5) 40
6) $4\sqrt{7}$
7) 3 9/37 hrs
8) 14
9) 1
10) 56
11) 25
12) 34
13) 25%
14) $\frac{2n-b}{3m-4n}$
15) $\frac{2}{m+n}$

Test - 19
1) 72
2) {1,2,3,4,5,6,7,8,9}
3) 159.984
4) 17/9
5) 29
6) 8
7) 12
8) 6
9) $4^3 < 3^{15} < 4^{15}$
10) 32,768
11) $42
12) 14 sec
13) 15
14) 5
15) -3

Test - 20
1) 126
2) 360
3) 20
4) 61/144
5) 1121
6) {a, b}
7) 650/3
8) 26
9) 30
10) $\frac{2}{3} < \frac{3}{4} < \frac{5}{6} < \frac{7}{8}$
11) 49
12) 170
13) 5
14) $\sqrt{30} + 5\sqrt{2}$
15) 150

Test - 21
1) 14/3
2) 560
3) 8
4) 10
5) {-3,3/2}
6) $\sqrt{17} + \sqrt{10}$
7) 22
8) 100
9) -9
10) 13/42
11) {-9/2,3/2}
12) $6\pi - 12$
13) $(3\sqrt{10}+5)/5$
14) 5:14
15) 20

Test - 22
1) 3025
2) 19
3) 16 2/3%
4) 1% decrease
5) 5
6) 10
7) 169/4
8) 4
9) $b < a < c$
10) $8 - 3x$
11) 90°
12) 5
13) -2
14) 12
15) 280

Test - 23
1) -10
2) 42
3) 123/146
4) 12/169
5) 18
6) $10a$
7) 20/7
8) 2
9) 84
10) 7/5
11) 1
12) 40.04
13) 11.9
14) 587/60
15) $m^2 + 2$

Test - 24
1) 144
2) 200
3) 2.25%
4) 306.25
5) 51
6) 37.5%
7) $.\overline{6}$ or 2/3
8) 332
9) 449/170
10) 1
11) 52/365
12) {-3/2,-1}
13) 180
14) 32/31
15) 19

Test - 25
1) 256
2) 196
3) 38
4) 360
5) 3,000,000
6) 5,000,000
7) 3
8) 10
9) 10
10) 14
11) 0
12) $m^3 + 49m$
13) 2^{22}
14) 31/16
15) 13

Test - 26
1) 3
2) $a^5 b^4 c^4$
3) 33
4) 48
5) 33
6) 8
7) $a^9 b^{10}$
8) $16a^{14} b^{12} c^{16}$
9) 121
10) k
11) $1382.40
12) 5 5/11
13) 1
14) 16
15) 2

Test - 27
1) 57 1/7
2) 42 6/7%
3) 314
4) {-1,-4/3}
5) $\frac{1}{7^5 \cdot 10}$
6) 5.1
7) 67/5
8) $abc(4a + 13bc)$
9) $2x + 1$
10) $x + 1$
11) $4b + \frac{ca}{3} + 1$
12) $\frac{4ab^2}{3c}$
13) -27/11
14) 12
15) 861

Test - 28
1) 4
2) 1
3) $\frac{5x+11}{6y}$
4) 3
5) 125
6) 484
7) 237.5 ltrs
8) 40
9) 1728
10) $56
11) 50
12) 54/67
13) 2/3
14) -30
15) $x^2 + 7x + 14$

Test - 29
1) $\frac{1}{2} < \frac{2}{3} < \frac{3}{4} < \frac{5}{6}$
2) 36
3) 15
4) 529
5) -17/9
6) 4
7) 53/55
8) $40
9) 7 1/7
10) 5/9
11) $b < a < c$
12) $b < a < c$
13) 3/2
14) 44
15) 3^{250}

Test - 30
1) 10
2) 16/11
3) $x < y < z$
4) 20 cm
5) 24/7
6) 16
7) 1067
8) 36
9) 34/3
10) $1 + \sqrt{2} + 3\sqrt{6}$
11) 10
12) 3
13) 64
14) $4x - 7$
15) $2\sqrt{10}$

Test - 31
1) 60°
2) 135°
3) 72°
4) 22.5°
5) 36°, 54°
6) 54°
7) 33°
8) 63°
9) 36°
10) 36°
11) 30°
12) 42°
13) 93°
14) 84°
15) 88°

Test - 32
1) $11\sqrt{2}/2$
2) $21+7\sqrt{3}$
3) $16\sqrt{3}$
4) 48/5
5) 12
6) 35
7) 16
8) 6/7
9) 9/2
10) $6\sqrt{3}$
11) 7
12) $2\sqrt{3}$
13) $9\sqrt{3}$
14) 60
15) $2\sqrt{3}$

Test - 33
1) 15
2) 27
3) 4 2/3 cm
4) $121\sqrt{3}/4$
5) 72
6) $12\sqrt{2}$
7) $49^x/16$
8) $P = 4 \cdot 7^x$,
   $A = 49^x$
9) 24
10) 4 cm
11) $8\sqrt{11}$
12) 192
13) $2\sqrt{41}$ cm
14) 300 cm²
15) 68°

Test - 34
1) 80
2) 320
3) 84
4) 16.5 cm²
5) 66
6) 15
7) 225 cm²
8) $72\sqrt{5}$ cm²
9) 100°
10) 280.8 cm²
11) 34°
12) 36°
13) 6π cm²
14) 88°
15) 38°

Test - 35
1) 40°
2) 90°
3) -3/2
4) 18 cm²
5) .75 cm
6) 108°
7) 80°
8) $25\sqrt{3}$ cm²
9) $3 < x < 13$
10) 60 cm
11) 6^x
12) $10\sqrt{2} + 10$ cm
13) 3
14) 1/2
15) $\sqrt{74}$

Test - 36
1) 35^x
2) 20
3) $\sqrt{3}/4$
4) $y = 2x - 2$
5) 1
6) 32
7) 6
8) $2(7^{2x} + 7^{3x})$
9) $3\sqrt{2}/4$
10) 216 cm²
11) 1
12) 2
13) 2/3
14) 6/7
15) -4/5

Test - 37
1) 42 cm²
2) 36 cm²
3) $2\sqrt{41}$ cm
4) 12π cm
5) 65π cm²
6) 196π cm²
7) $A = 38$ cm², $V = 12$ cm²
8) $A = 216$ cm², $V = 216$ cm³
9) 8 cm³
10) 150 cm²
11) 8/27
12) 49/25
13) 24 cm³
14) 5 cm
15) $A = 400\pi$ cm², $V = 4000\pi/3$ cm³

About the Authors

Dr. Steve Warner earned his Ph.D. at Rutgers University in Mathematics, and he currently works as an Associate Professor at Hofstra University. Dr. Warner has over 15 years of experience in general math tutoring and over 10 years of experience in SAT math tutoring. He has tutored students both individually and in group settings and has published several math prep books for the SAT, ACT and AP Calculus exams.

Tayyip Oral graduated from Qafqaz University in Azerbaijan in 1998 with a Bachelor's degree in Engineering, and he received an MBA from the same university in 2010. Tayyip is an educator who has written several books related to math and intelligence questions. He has previously taught math and IQ classes at Baku Araz preparatory school for 13 years.

BOOKS BY TAYYIP ORAL

BOOKS BY DR. STEVE WARNER

 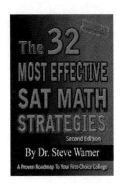  

  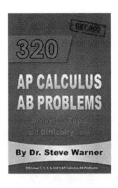 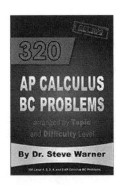

Made in the USA
San Bernardino, CA
20 November 2015